Amazing Animals

Multiplying Multidigit Numbers by One-Digit Numbers with Regrouping

Orli Zuravicky

Rosen Classroom Books & Materials
New York

Published in 2004 by The Rosen Publishing Group, Inc.
29 East 21st Street, New York, NY 10010

Copyright © 2004 by The Rosen Publishing Group, Inc.

All rights reserved. No part of this book may be reproduced in any form without permission in writing from the publisher, except by a reviewer.

Book Design: Ron A. Churley

Photo Credits: Cover © Gail Shumway/Taxi; cover (left), p. 12 (top) © PhotoDisc; cover (right) © Lynn Stone/Index Stock; p. 5 (frog) © David Tipling/The Image Bank; pp. 5 (snake), 9 (snake) © EyeWire; p. 5 (elephant) © Jonathan Scott/Taxi; p. 5 (panda) © David Thurston/Taxi; pp. 6 (bullfrog), 7 © Joe McDonald/Corbis; p. 6 (top) © G. K. & Vikki Hart/The Image Bank; p. 8 (left) © Alex Kerstitch/Taxi; p. 8 © Anthony Bannister/Corbis; p. 10 © Jeff Rotman/The Image Bank; p. 10 (left) © John Warden/Index Stock; p. 11 (bottom) © Jeffrey L. Rotman/Corbis; p. 11 (top) © Volvox/The Image Bank; p. 12 (bottom) © Francesc Muntada/Corbis; p. 13 © Zefa Visual Media-Germany/Index Stock; p. 14 © Beverly Joubert/National Geographic; p. 15 (bottom) © Martin Harvey; Gallo Images/Corbis; p. 15 (top) © Stephen Johnson/Stone; p. 16 (inset) © James Gritz/PhotoDisc; p. 17 © Volvox/Index Stock; p. 18 © Keren Su/Stone; p. 18 (inset) © Bill Lai/Index Stock; p. 19 (bottom) © Tom Davis/Stone; p. 19 (top) © AFP/Corbis; p. 20 (inset) © RO-MA Stock/Index Stock; p. 21 © Kevin Wardius/Index Stock; p. 22 © Matt W. Moffett/Minden Pictures.

Library of Congress Cataloging-in-Publication Data

Zuravicky, Orli.
   Amazing animals : multiplying multidigit numbers by one-digit numbers with regrouping / Orli Zuravicky.
       p. cm. — (PowerMath)
Includes index.
Summary: Gives facts about familiar animals, such as how often a snake sheds its skin and how many pounds a baby blue whale gains in a week, and shows how to use multiplication to determine further information.
   ISBN 0-8239-8964-X (lib. bdg.)
   ISBN 0-8239-8861-9 (pbk.)
   6-pack ISBN 0-8239-7345-X
  1. Multiplication—Juvenile literature. 2. Problem solving—Juvenile literature.
 [1. Multiplication. 2. Problem solving. 3. Animals—Miscellanea.] I. Title. II. Series.
   QA115.Z87 2004
   513.2'13—dc21
                              2002154590

Manufactured in the United States of America

Contents

The World of Multiplication   4

Frogs in Your Backyard   6

Skillful Snakes   9

Shark Teeth   11

A Spider's Silk   12

An Elephant's Dinner   15

Baby Whales   16

A Panda's Favorite Food   19

Wings That Hum   20

Math Is Everywhere!   22

Glossary   23

Index   24

The World of Multiplication

Math is an important part of everyday life. One of the most important math skills we use is multiplication. Some multiplication problems require regrouping. Sometimes when you are multiplying, the answer in a place—like the ones place—will have two **digits**. You have to regroup the answer so that there is just one digit in each place. This is something you have already learned to do in addition problems. We can use this skill to discover lots of fun facts about the amazing animals that live all around us.

Do the Math

$$\begin{array}{r} \overset{2}{3}4 \\ \times\ 5 \\ \hline 170 \end{array}$$

Regrouping happens in a multiplication problem when the answer for one place is a two-digit number. In this problem, 5 x 4 equals 20. The 0 stays in the ones place below the line, and the 2 gets carried up and placed over the 3. Then, 5 x 3 equals 15, but your work isn't done until you add the 2 to the tens place.

Frogs in Your Backyard

Have you ever seen a frog in your backyard? Frogs are **amphibians**. They live half of their lives in water and half on land. The bullfrog is the largest frog found in America. A bullfrog can leap 12 times its own body length in 1 jump.

How far can an 8-inch bullfrog jump in 5 leaps? First, let's figure out how far it can jump in one leap. Multiply its length, 8 inches, by 12 to get 96 inches. To find out how far it jumps in 5 leaps, multiply the length of one jump, 96 inches, by 5. This bullfrog can jump 480 inches in 5 leaps!

bullfrog

Do the Math

```
    1
   12
 x  8 inches
   96 inches
```

```
      3
    96 inches
 x   5 leaps
   480 inches
```

Regrouping is necessary when multiplying 8 x 2 in the first problem. The answer, 16, is a two-digit number. Place the 6 below the line. Carry the 1 up and place it above the number in the tens place, 1. Now multiply 8 by 1, which equals 8. However, your answer isn't final until you add the 1 you carried over to the 8. Then 8 + 1 equals 9. Your final answer is 96 inches.

Skillful Snakes

Over 2,000 different kinds of snakes live around the world. Snakes are **reptiles**. They are **cold-blooded** and have scaly skin. When a snake outgrows its skin, it **molts**, or sheds it. Some **tropical** snakes molt 6 times a year.

In 15 years, about how many times would one of these tropical snakes have molted? Multiply 15 years by the number of times it molted each year, 6. This snake molted about 90 times in its life.

```
  3
 15 years
x  6 times
 90 times
```

Do the Math

```
  1
 13 years
x  6 times
 78 times
```

If this snake lived to be only 13 years old, how many times would it have molted in its life? To find out, multiply 13 years by 6. The answer is 78 times.

Do the Math

$$\begin{array}{r} \overset{5\,1}{172} \text{ teeth} \\ \times \quad 8 \text{ days} \\ \hline 1{,}376 \text{ days} \end{array}$$

Multiplying the number of teeth lost (172) by the days needed to replace each tooth (8) will give you the right answer. It would take 1,376 days for this shark to replace 172 teeth. That's almost 4 years!

Shark Teeth

Did you know that sharks were around even before the dinosaurs? The earliest sharks appeared on Earth about 400 million years ago. There are about 370 different kinds of sharks, which live in waters all over the world. Some sharks are as small as 6 inches. The largest sharks can be as long as 50 feet!

Sharks' mouths are filled with rows of teeth. When a shark loses a tooth, it is replaced by the tooth behind it. Lemon sharks, named for their yellow back, can replace 1 tooth in just 8 days. How many days would it take for a lemon shark to replace 172 teeth?

lemon shark

A Spider's Silk

Many people think that spiders are **insects**, but they aren't. Spiders belong to a group of animals called **arachnids**. Spiders have only 2 body parts, while insects have 3. Most insects have 6 legs, but spiders have 8.

Most spiders build webs with silk they make in their bodies. Different spiders make different kinds of silk and build webs of different shapes. Spider silk is the strongest natural fiber that exists. It is even stronger than steel! Spider silk can be **stretched** up to 2 times its length without breaking.

Do the Math

$$\begin{array}{r} \overset{1}{}29 \text{ inches} \\ \underline{\times\ 2 \text{ times}} \\ 58 \text{ inches} \end{array}$$

If a strand of spider silk is 29 inches long, how far could it stretch without breaking? Since it can stretch 2 times its original length without breaking, multiply 29 inches by 2. It can stretch up to 58 inches without breaking!

Do the Math

$$\begin{array}{r} \overset{4}{}\\ 16 \text{ hours}\\ \times\ 7 \text{ days}\\ \hline 112 \text{ hours} \end{array}$$

If a wild elephant spends 16 hours eating in one day, how many hours will it spend eating in one week? Just multiply 16 hours by 7 days. That's right, an elephant spends 112 hours a week eating. That's about $4\frac{1}{2}$ days!

An Elephant's Dinner

Elephants are the largest animals that live on land. They live in the wild in Africa and Asia. You might have seen elephants in a zoo. Elephants are the second tallest animals in the world and can weigh more than 12,000 pounds.

Elephants have long trunks that they use to smell and breathe. An elephant's trunk also acts as an arm, which they use to pick up their dinner. Elephants eat grass, plants, leaves, and fruits. A wild elephant eats about 300 pounds of food a day and spends 16 hours each day eating.

Baby Whales

The blue whale is the largest animal that has ever lived, either on land or in the water. A blue whale can be over 100 feet long and can weigh 300,000 pounds. Whales are **mammals**, just like people are. They are **warm-blooded**, they give birth to live babies, and mothers feed their young with milk from their bodies. Blue whales give birth to one baby at a time. Baby blue whales are called calves. When a calf is born, it can be 23 feet long and weigh as much as 4,000 pounds. A baby blue whale can gain up to 250 pounds a day.

Do the Math

$$\begin{array}{r} \overset{3}{}250 \text{ pounds} \\ \times\ \underline{7} \text{ days} \\ 1{,}750 \text{ pounds} \end{array}$$

How many pounds can a baby blue whale gain in 1 week? Since there are 7 days in a week, you simply multiply the amount it can gain in 1 day (250 pounds) by the number of days (7). A baby blue whale can gain up to 1,750 pounds in 1 week!

bamboo

Do the Math

```
  2
  85  pounds of bamboo
x  5  pandas
 425  pounds of bamboo

          2
          85  pounds of bamboo
        x  5  days
         425  pounds of bamboo
```

If each panda eats 85 pounds of bamboo a day, how many pounds of bamboo do 5 pandas eat in 1 day? How many pounds of bamboo does 1 panda eat in 5 days? If you did the math correctly, you should get the same answer.

A Panda's Favorite Food

Pandas are giant black and white animals that live in Asia. Some scientists think that pandas are bears. Other scientists say that they aren't because bears eat other animals and pandas do not.

Pandas eat mainly bamboo. Bamboo is a giant grass that grows mostly in tropical areas. Even though pandas eat plants, their bodies have trouble changing plants into energy. They have to eat a lot of bamboo in order to get enough energy to live. A giant panda eats about 85 pounds of bamboo a day.

Wings That Hum

Hummingbirds are colorful birds with long, thin bills for sipping the sweet liquid found inside flowers. The bee hummingbird is the smallest bird in the world. It is only 2 inches long. The giant hummingbird is just 8 inches long.

Hummingbirds have the fastest wings in the world. In fact, their name comes from the humming sound their wings make. Scientists have measured a ruby-throated hummingbird's wing speed at 75 beats per second! How many times would this bird's wings beat in 3 seconds?

Do the Math

```
  1
 75 beats per second
x 3 seconds
─────
225 beats
```

Do the Math

```
    3
  75  beats per second
 x 6  seconds
 450  beats
```

If a ruby-throated hummingbird beats its wings 225 times in 3 seconds, how many times would it beat its wings in 6 seconds? To find out, multiply 75 by 6. You could also multiply 225 by 2. The answer is 450 times.

```
    1
 225  beats
  x 2 times
 450  beats
```

Math Is Everywhere!

There are many fun new facts you can learn by using math skills. Your friends will be amazed by all the facts you've figured out, like how much weight a baby blue whale gains in 1 week and how far a bullfrog can jump. Once you've figured out all the problems in this book, try to come up with your own. The world we live in is full of math problems. If we take the time to understand and practice multiplication and regrouping, we can discover new things about the world we live in.

Glossary

amphibian (am-FIH-bee-uhn) A cold-blooded animal that has moist skin. Amphibians live part of their lives in water and part on land.

arachnid (uh-RAK-nid) A small animal that has two body parts, a hard outer covering, and no bones. Spiders and scorpions are arachnids.

cold-blooded (KOLD–BLUH-dud) An animal that has the same body temperature as the surrounding air.

digit (DIHJ-it) Any of the figures 0, 1, 2, 3, 4, 5, 6, 7, 8, and 9.

insect (IN-sekt) A small animal with three body parts, six legs, a hard outer covering, and no bones.

mammal (MA-muhl) A warm-blooded animal that is often covered with hair or fur. Mothers feed their babies milk from their bodies.

molt (MOHLT) To shed an outer covering like skin, scales, fur, or feathers.

reptile (REP-tyle) A cold-blooded animal that has dry, scaly skin and lays eggs.

stretch (STRECH) To pull something so hard that it becomes longer.

tropical (TRAH-pih-kuhl) Found in areas where it is very warm all year.

warm-blooded (WARM–BLUH-dud) An animal that has the same body temperature no matter what the outside temperature is.

Index

A
Africa, 15
amphibians, 6
arachnids, 12
Asia, 15, 19

B
bamboo, 19
blue whale(s), 16, 22
bullfrog, 6, 22

C
cold-blooded, 9

E
elephant(s), 15

H
hummingbird(s), 20

I
insects, 12

L
leap(s), 6
lemon shark(s), 11

M
mammals, 16
molt(ed), 9

P
panda(s), 19

R
regroup(ing), 4, 22
reptiles, 9

S
shark(s), 11
silk, 12
snake(s), 9
spider(s), 12

T
teeth, 11

W
warm-blooded, 16
webs, 12
wing(s), 20